Armadillos Sleep in Dugouts

and other places animals live

Pam Muñoz Ryan

Illustrated by Diane deGroat

Hyperion Books for Children
New York

Gray squirrels make a cozy drey
from moss and twigs and leaves.

River otters live on a holt
among the shrubs and trees.

Honeybees build a beehive
with rooms of honeycomb.

I'm a bustling beaver.
Where do I call home?

Beaver, you are a builder,
and although it seems hodgepodge,
your home of branches, sticks, and mud
becomes a sturdy lodge.

Moose stamp out a moose yard
when it's time to take a rest.

A polar bear builds a snow cave,
where she and her cubs will nest.

Walruses take a slumber
on ice floes in the sea.

I'm an Arctic lemming.
Where will my household be?

Lemming, you live in tundra grass
until winter blizzards blow.
Then you live inside a runway
between the earth and snow.

Foxes live in cozy dens
where they can hide away.

Hares scrape out a shallow form
and doze throughout the day.

Moles live in a citadel
with tunnels underground.

I'm a peregrine falcon.
Where is my front door found?

Falcon, you fly to a place up high
to nest or rest or hide.
You roost in a lofty aerie
in a tree or mountainside.

Badgers live in rambling setts
in furrows that connect.

A pack rat builds a hut of sticks
for the things it will collect.

A spider spins a web and waits
for dinner to come along.

I'm a desert cactus wren.
Where do I belong?

Wren, you choose a cholla cactus
with prickly thorns and spines.
It's your condominium
during desert rain or shine!

African termites build a dirt mound
with chimneys tall and steep.

Leopards find a protected lair
when it's time to sleep.

Long-eared hedgehogs dig a burrow
for their sleeping space.

I'm a hippopotamus.
Do I have a favorite place?

Hippo, you like a shallow river
for your water bed.
You snooze in a murky wallow
showing nothing but your head.

Armadillos sleep in dugouts
covered up with vegetation.

Pampas deer find a thicket
when they want a safe location.

Rufous ovenbirds build an 'oven'
when it's time to raise their brood.

I am a three-toed sloth.
Where do I live and get my food?

Sloth, you like a leafy house
that is far above the ground.
You live upon a tree branch
while dangling upside down!

Mountain possums like a hollow stump
where several may get together.

Skinks crawl in a crevice
to protect them from the weather.

Birdwing caterpillars spin a snug cocoon
when it's time to change and grow.

I am a gardener bowerbird.
Do I have a place to go?

Bowerbird, you're quite creative
with berries, shells, and flowers.
When it's time to attract a mate,
you build a courtship bower!

Whether den or nest or tree house
or an apartment underground,
most animals find a certain spot
where they feel safe and sound.

But I don't seem to have a home
or a special place to be.
I am an Asian elephant.
What about ME?

Elephant, you don't *have*
a sheltered house.
You wouldn't fit inside!

You don't have a hideaway.
You're far too big to hide!

You hang out in the great outdoors,
where you are free to roam,
and *wherever* you happen to stop
for a while . . .

. . . is the place that you call home.

To my mom and dad, Don and Hope, who made their house a home.

—P. M. R.

Other books by the same author:
One Hundred Is a Family
and
A Pinky Is a Baby Mouse

Armadillos Sleep in Dugouts by Pam Muñoz Ryan,
illustrated by Diane deGroat
Text © 1997 by Pam Muñoz Ryan.
Illustrations © 1997 by Diane deGroat
Reprinted by permission of Hyperion Books for children.

Printed in the United States of America

The artwork for each picture is prepared using watercolor.
This book is set in 15-point Stempel Schneidler.

Library of Congress Cataloging-in-Publication Data
Ryan, Pam Muñoz.
Armadillos sleep in dugouts and other places animals live / Pam Muñoz Ryan; illustrated by Diane deGroat.–1st ed.
p.   cm.
Summary: Examines the different types of homes animals make, including those of river otters, peregrine falcons, and three-toed sloths.
ISBN 0-7868-0274-X (trade)—ISBN 0-7868-2222-8 (lib. bdg.)
1. Animals—Habitations—Juvenile literature. [1. Animals— Habitations.] I. deGroat, Diane, ill. II. Title.
QL756.R93   1997
591.56′4—dc20   96-35931